FROM LIVING CELLS TO DINOSAURS

FROM LIVING CELLS TO DINOSAURS

ROY A. GALLANT

ILLUSTRATIONS BY
ANNE CANEVARI GREEN

A FIRST BOOK
FRANKLIN WATTS 1986
NEW YORK • LONDON
TORONTO • SYDNEY

FOR
JENNIFER

FRONTIS: HUGE, PEACE-LOVING, PLANT-EATING
DUCK-BILLED DINOSAURS ROAMED THE MESOZOIC PLAINS.

Photographs courtesy of: Department of Library Services, American Museum of Natural History: pp. 2 (#335200, R.E. Logan), 11 (#110825, Rice), 23 (#318465, Charles H. Coles), 25 (#125632), 32 (#34823, J. Kirscher), 37 (#318274, T.L. Bienwert), 42 (#322872), 48 (#242201), 52 (#110010, J. Kirscher), 56 (#45558, E.O. Hovey), 59 (#2A3384), 70 (#123825, Logan), 71 (#35799), 73 (#34711), 78 (#318857, Charles H. Coles), 84 (#122115); The Bettmann Archive: p. 18; Field Museum of Natural History, Chicago: pp. 26, 29, 35, 41, 45, 47 (Charles R. Knight, artist), 80–81 (Charles R. Knight, artist), 83 (Charles R. Knight, artist); Arizona Development Board: p. 55; U.S. Geological Survey: p. 61; U.S. Forest Service: p. 66; AP/Wide World: p. 68.

Library of Congress Cataloging-in-Publication Data

Gallant, Roy A.
From living cells to dinosaurs.

(A First book)
Includes index.
Summary: Discusses the changes that took place on land and in plants and animals, from the Precambrian to the Mesozoic era, when dinosaurs reached their peak.
1. Paleontology—Juvenile literature. (1. Paleontology) I. Green, Anne Canevari, ill. II. Title.
QE714.5.G37 1986   560   86-11111
ISBN 0-531-10207-6

Copyright © 1986 by Roy A. Gallant
All rights reserved
Printed in the United States of America
5 4 3 2 1

CONTENTS

CHAPTER ONE
Life Establishes
Itself on Earth
9

CHAPTER TWO
The Diversity of Life
16

CHAPTER THREE
From Sponges
to Fishes
22

CHAPTER FOUR
From Fishes to
Amphibians to Reptiles
34

CHAPTER FIVE
Life During the Triassic Period
51

CHAPTER SIX
Life During the Jurassic Period
63

CHAPTER SEVEN
Life During the Cretaceous Period
75

Glossary
89

Index
94

ACKNOWLEDGMENTS

I wish to thank Dr. George Kukla, of the Lamont-Doherty Geological Observatory, Palisades, New Jersey, for his thorough reading of this book in its manuscript stage to check for accuracy and for his suggestions for making certain passages more suitable for young readers.

My thanks also to my editor, Maury Solomon, for her customary thoroughness in keeping the text at a suitable reading level and for her good sense in reorganizing certain sections of the book.

# THE EARTH IN TIME
Numbers = millions of years ago

MESOZOIC ERA "Age of Reptiles"

190 M. YEARS AGO

PALEOZOIC ERA "Ancient Life"

215 — PERMIAN — 155 TRIASSIC — 120 JURASSIC — CRETACEOUS — 70 MILLION YEARS AGO

500 CAMBRIAN 440

PENNSYLVANIAN 235 MISSISSIPPIAN ("Age of Amphibians") 265 DEVONIAN ("Age of Fishes") 330 SILURIAN 365 ORDOVICIAN

PRECAMBRIAN ERA

CENOZOIC ERA "Recent Life"

PROTEROZOIC

TERTIARY

1.8 QUATERNARY

ARCHAEOZOIC

4.5 billion years ago

CHAPTER ONE

LIFE ESTABLISHES ITSELF ON EARTH

FOSSIL BRIDGES TO THE PAST

As we read the fossil record—that is, the record of life as it was in past ages—we are drawn back through time—back 100 million years to a time when inland seas covered most of Europe and nearly half of North America; back 180 million years when reptiles, including the dinosaurs, ruled the land and when birds and mammals first appeared; back 300 million years, when vast swamps sprawled over much of the Northern Hemisphere and when amphibians ruled the land; back 380 million years to the Age of Fishes and before there were any land animals; back 550 million years to when the Cambrian seas abounded with trilobites, brachiopods, and other animals without backbones; and finally back into the dark of Precambrian times.

The geologic time span we will be looking at in these first few chapters is the Precambrian, which stretches from 4.6

billion years ago—the estimated age of Earth—to 570 million years ago and covers 7/8 of Earth's history.

Fossils show that near the end of the Precambrian there were jellyfishes, corals, and burrowing worms. All the organisms living then had soft bodies. Hard-bodied animals with shells and sharp protective spines were not to evolve until later.

We can view the Precambrian as the time when life was firmly establishing itself on Earth. It had not yet begun to show the splendid variety that was to come later.

The deeper we probe into earlier and earlier times of the Precambrian, the simpler and more primitive the fossils become. About thirty years ago paleontologists discovered tiny fossils in sedimentary rock in Fig Tree, South Africa. They are about 3.2 billion years old. In the 1970s a geological expedition searching near the town of North Pole in Western Australia found dome-shaped rocks made up of hundreds of wafer-thin layers, called stromatolites. These little mounds are the fossil remains of colonies of primitive bacteria. They are the oldest-known fossils and are about 3.5 billion years old.

Were there any living organisms before that? Most likely yes. To say what events led to the formation of those earliest known simple biological cells, scientists had to do a lot of guesswork. They had to use their knowledge of how certain simple chemicals of life join and form more and more complex chemicals.

FROM CHEMICALS TO "LIVING" MATTER

The bodies of all plants and animals are made up of many chemicals. For life as we know it to have started on Earth, the planet must have had at least twenty major chemicals. The elements hydrogen, carbon, nitrogen, oxygen, phosphorus,

Corals appeared near the end of the Precambrian period. Shown here is fossilized coral.

PRECAMBRIAN FOSSILS

AGE	KIND AND LOCATION
900 million years	Microscopic plants found in Bitter Springs, Australia. Thirty kinds have been found, 14 of which are like certain green plants, called algae, living today.
1.6 to 2.0 billion years	Microscopic algae found in rock along the northern shore of Lake Superior in Ontario. Twelve kinds have been found. Bacteria, in the shape of rods, also have been found.
3.1 billion years	Microscopic bacterialike organisms found near Fig Tree, South Africa. Similar to blue-green algae today.
3.4 to 3.6 billion years	Microscopic organisms like those found near Fig Tree have been found in Swaziland, Africa.

and sulfur make up about 95 percent of our bodies and the bodies of most other living animals and plants. Other elements needed in lesser amounts include potassium, sodium, magnesium, calcium, and chlorine.

So far as we can tell, life arose in the early warm seas. Chemists have shown that energy from the Sun, from lightning, and from other sources can cause such common elements as carbon, oxygen, nitrogen, and hydrogen to join and form the complex molecules needed for life. Among these molecules are certain chemical building blocks called amino acids. Amino acids combine with each other to form still larger building blocks of life called proteins. Our bodies use proteins for growth, repair of injured parts, and as a source of energy.

In recent years astronomers have discovered that certain life-giving chemicals exist in space and in meteorites. Some astronomers wonder if certain key molecules important to life arrived in meteorites early in our planet's history. In September 1969 a meteorite that fell in southern Australia contained several different kinds of amino acids.

But how could nonliving molecules, no matter how complex, become living matter?

Here we may be setting a word trap for ourselves by thinking of things that are living and other things that are not living, with nothing in between. Actually we should be thinking of an unbroken chain of chemical forms ranging from elements to simple molecules to complex clumps of molecules, and eventually to systems of molecules that have certain things in common with matter that we say is "living" matter.

If indeed life developed in this manner, the process must have taken a long time, perhaps a billion years. The length of such a time span is almost impossible for the human mind to grasp. If we let one year be represented by the thickness of a page of this book, then one billion years would be represented by a stack of books (without their covers) 50 miles (80 km) high.

The next stage in the evolution of living matter might have been complex groupings of large molecules into microscopic units enclosed within protective jackets, called membranes. Such clumps of living matter were early forms of protein. The membrane of each cluster acted as a sac that separated the outside environment from the environment inside the sac. However, tiny holes in the sac let certain food molecules from the outside enter the sac. The food molecules were used as building blocks for growth and as a source of energy. The holes also let waste matter made by the proteinlike cluster escape to the outside. This is the situation suggested by our knowledge about living membranes today.

We can imagine more and more of these proteinlike clumps of living matter, enclosed within a protective membrane, forming and taking in food molecules from the outside. We can also imagine certain of these proteinlike clumps becoming advanced in a way that gave them an advantage. Instead of taking in existing food molecules, they took in other, smaller molecules and then assembled these into food molecules. Since there would be more of the smaller molecules in the outside environment than ready-made food molecules, those proteinlike clumps able to make their own food would have an advantage over those that had to find their food ready-made.

In some such process, the first living cells evolved. Cells are the smallest organized units of living matter recognized by biologists. Your body is made up of billions of cells. Cells were organisms that became expert at making their own food out of raw materials in the outside environment. The raw materials were carbon dioxide molecules and water vapor. With sunlight as a source of energy the cells combined the carbon dioxide and water vapor and made a sugar called glucose. In the process they gave off oxygen as "waste" matter to the air. These are the organisms that might have evolved as early as some 3.7 billion years ago.

At first the oxygen freed by those primitive organisms, whatever they were like, was quickly absorbed by iron and certain other oxygen-hungry elements in the early seas. Not until these elements of the seas had satisfied their chemical appetite would the oxygen given off by those organisms enter the air and remain there as "free" oxygen.

There are many gaps in our knowledge of the chain of chemical events that led to the first cells. Even so, most biologists agree that some such broad outline like the one just described seems likely. Biologists also generally agree that with the right conditions—that is, the right blend of matter and energy—life of some sort is bound to arise due to natural causes. This is so not only on Earth but on the countless other planets we believe circle the countless number of stars visible to us. Although we do not yet have any direct evidence of life elsewhere in the universe, the chances that it does exist seem overwhelming.

Closer to home, today we know that Earth is the only planet circling the Sun able to support the many forms of life we see around us. But simple or unique life forms may one day be found in the soil of Mars or within the atmosphere of Jupiter. Right now we just don't know.

CHAPTER TWO

THE DIVERSITY OF LIFE

If we take the time to look about us carefully, we cannot help but be impressed by the many different kinds of plants and animals, or, as biologists say, by the diversity of life. About 1,200,000 different kinds, or species, of animals have been listed in catalogs, and about 500,000 species of plants. And each year another 15,000 or so species are added to the list. This may seem to be a surprising diversity of life forms. But even more surprising is the fact that more than 99 percent of all the species that have ever lived on Earth are now extinct!

    Life invades every available nook and cranny on our planet. There are bacteria that live in hot springs with temperatures above the boiling point. There are plants that live on mountaintops where the temperature is always below freezing. There are air-breathing mammals, such as whales, that live under water. There are ants and spiders and people and lobsters, and there are birds galore. All of these life forms evolved from those first living cells.

HOW ADAPTATION WORKS

A herd of cows, a colony of ants, a flock of geese, or any other healthy population is suited to, or adapted to, its environment. A population of desert animals, for example, is adapted to the heat and dryness of the desert. The population would die if moved to the Arctic. Likewise, a population of polar bears could not survive in a desert environment.

But over time the environment changes. Ice ages come and go. New mountains are thrust up and change the local climate. Swamps and inland seas dry up. Glaciers melt. All such changes force new conditions on the populations of plants and animals living there. Sometimes such changes are so severe that no individual members of the population can survive, and the population dies out. In other cases, when the change is less severe, certain individuals that are fitter than the others are able to survive. For example, a squirrel with a thicker coat of hair might be better able to survive a series of especially cold winters than a squirrel with only a thin coat. A moth with wing colors that make it especially hard for birds to find is better protected, or adapted, than moths that are less well camouflaged. Such "favored" individuals in a population appear to be produced randomly (by chance) in a process called mutation.

Mutation is a change in an animal's genes that make the animal different in one or more ways from its parents. For example, one puppy in a litter may have one brown eye and one blue eye, while all its brothers and sisters have both eyes the same color. X rays and certain other forms of radiation and pollution can cause such mutations. Although most mutations are harmful, once in a while one just happens to

Charles Darwin caused a great stir when he published his book On the Origin of Species *outlining his theory of evolution.*

WHAT IS EVOLUTION?

Animals or plants of the same kind are said to belong to a particular species. For example, dogs belong to one species, human beings to a different species, and white pine trees to still a different species. The fossil record shows that species arise and die out. After living for more than 100 million years, the dinosaurs as a group died out about 65 million years ago. But the dinosaurs had not always existed, just as people, roses, and termites have not always existed. The fossil record further clearly shows that species change, or evolve. The general process of a species changing is called evolution. Two naturalists, Charles Darwin and Alfred Russell Wallace, are credited with developing the theory of evolution, which has become the unifying principle of all biology. Here are Darwin's major ideas about evolution:

1. The individuals making up a population, whether it is a human population, a population of squirrels, or a population of birch trees are not all exactly alike. Among the human population, for example, there are different races, some individuals that are better able to fight off disease than others, some that have better eyesight than others, and so on. Such differences are called variations.

2. Some variations, such as the ability to fight off certain diseases, are helpful and may be passed on from parents to their children.

3. Some of the children born into a population have certain inborn advantages over certain other children who may not be able to fight off disease very well, for example. The healthier children have a better chance to live longer than the children who are not so healthy. Such "favored" individuals tend to have a better chance of becoming parents and passing their good traits along to their children.

4. The result of such naturally favored individuals producing more children than unhealthy individuals do is an increased number of healthy individuals in the next generation.

5. So nature selects *for* those individuals of any population of animals or plants who are fittest and *against* those individuals who are less fit. That is what is meant by the expression "survival of the fittest." In this way each new generation is a little different from the previous one. It is those differences from one generation to the next that cause any species to change, or evolve.

The millions of species of animals and plants that have shared the Earth all evolved from the first primitive cells by a natural process that took hundreds of millions of years. This truth is beautifully revealed in the fossil record of past life.

turn out to be beneficial, as in the case of the squirrel with an especially thick coat of hair. When that happens we say that the individual with the mutation is *well adapted* to the new environment. Those individuals that are well adapted and survive may then pass on their special adaptation to their offspring. Those without the new adaptation are doomed to die before reproducing. Gradually the population once again grows in number as its new and fitter individuals produce offspring with the beneficial mutation. Those offspring, like their parents, are different from those who were unable to survive the environmental change. This is basically how the process called evolution changes populations and produces new species of plants and animals.

As it has in the past, biological evolution continues today. Its pace is too slow for us to see it happening directly. When we examine the fossil record, however, patterns of evolution become clear to us. Evolution itself is no longer just a theory. It is a fact accepted by almost all scientists. But there are different theories that try to account for the ways in which evolution takes place.

We could spend much more time describing the biological events thought to have taken place during the Precambrian era. But we have looked at the major ones. We will now continue our role as detective-scientists and find out what the fossil and rock records tell us about the next 325 million years of the Earth's history. That time span is called the Paleozoic era and begins with an explosive diversity of life into many new plant and animal forms.

CHAPTER THREE

FROM SPONGES TO FISHES

In the next two chapters we will explore what scientists think happened to the land and the seas, and how life evolved during the 325-million-year geologic time span known as the Paleozoic era. The name of this era comes from the Greek words *palaios,* meaning "ancient," and *zoe,* meaning "life." The Paleozoic is made up of seven shorter geologic periods. The three we will explore in this chapter are the Cambrian, the Ordovician, and the Silurian periods.

CAMBRIAN PERIOD

The Cambrian is named after Cambria, the Roman name for Wales in Great Britain. This geological period spans 85 million years and lasted from about 590 million years ago to 505 million years ago.

Many new life forms had appeared on the scene by the beginning of the Cambrian. Cambrian fossils are found in

Cambrian marine landscape

Wales and many other parts of the world, including the eastern United States and Canada.

The early Cambrian climate seems to have consisted of long cold spells that later were to be followed by warmer spells. Early in this period the central region of North America was dry land; but later on sinking of the land allowed shallow inland seas to form. By the end of the period almost all of the continent had been flooded.

These shallow seas abounded with many forms of life, including sponges, trilobites, brachiopods, graptolites, and other animals lacking backbones. Such animals without backbones are called invertebrates. At this stage in Earth's history neither plants nor animals had yet taken up life on the land. And animals with backbones had not yet evolved.

Trilobites were especially plentiful. There were thousands of species, the largest being about the size of a serving platter but most being only 2 inches (5 cm) long. Some were swimmers, others walked or burrowed in the mud. Some had efficient eyes, others lacked eyes altogether. All had several pairs of hard, jointed legs and lacked jaws and teeth. Their boneless bodies were protected by a tough shell. As the animal grew too large for its shell, it shed it and grew a new and larger one. These organisms were indirect ancestors of crabs, shrimps, spiders, scorpions, and insects, which were to evolve later. Judging by the diversity of trilobite species at the beginning of the Cambrian, these organisms must have evolved much earlier, during the Precambrian.

Such widespread diversity among the animals of the warm, late-Cambrian seas meant competition among species. And competition includes animals that eat other animals. Those that were most successful in the survival game were the ones who had developed hard shells, among them the trilobites. The trilobites make up about 70 percent of the Cambrian fossil record.

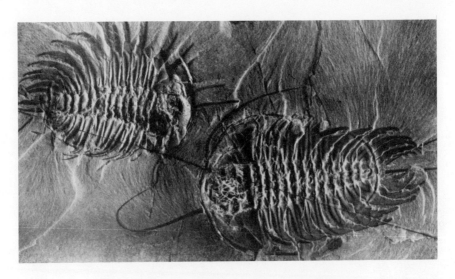

Trilobites make up about 70 percent of the Cambrian's fossil record.

One of our best windows into the Cambrian world is in the Canadian Rockies at Burgess Pass, Mount Wapta. Sedimentary rock (rock made from loose sediments such as clay and sand) containing many Cambrian fossils exists there in a deposit known as the Burgess Shale. In addition to containing the hard-bodied fossils of trilobites, the rock also contains much rarer remains of delicate sea organisms, including jellyfishes and sponges.

During the Cambrian in North America, Europe, and Asia, long ditches on the continents were filling up with sand, clay, and other sediments. Much later these sediment dumps were to be thrust up as mountain ranges. So it was during the Cambrian that our Rocky Mountains and the Appalachians, for instance, were beginning to form.

Ordovician landscape

ORDOVICIAN PERIOD

The Ordovician was named after an ancient Celtic tribe, the Ordovices, living in Wales. This geologic period spans about 65 million years and lasted from about 505 million years ago to about 440 million years ago.

The Ordovician continued as a period of rich marine life. Trilobites reached their greatest numbers, as did bryozoans, brachiopods, graptolites, and crinoids. Many bryozoans thrive in today's seas and include mosslike organisms, others shaped like fans, and still others in the form of thick stems. Bryozoans are small animals that live attached to the sea bottom. Around their mouth are little tentacles used to capture food that drifts by. Bryozoans lack bones but have a hard outer skeleton.

Brachiopods named *Lingula* thrive to this day in mudflats around the world and are almost exactly like their Cambrian ancestors. However, Cambrian brachiopods lived only in the seas and had two shells, like a clam. They lived attached to the bottom and with small hairlike tentacles captured food that drifted along.

Mollusks included shelled animals like today's clams and oysters, and cephalopods, which were highly successful during this period. The name *cephalopod* means "head-foot." Some cephalopods lived in coiled shells; others lived in long, cone-shaped shells. When young, a cephalopod lived in a small, simple shell chamber. But as it would outgrow its shell chamber it would form a new and larger one. So it kept adding new chambers and growing into them until the train of chambers became quite long. In all, there have been about ten thousand species of cephalopods. They are the most highly developed of all the mollusks. About 400 species are alive today, octopuses and squids among them. Most, but not all, mollusks have hard shells of one or several sections. They also have a digestive tract with a front and rear opening, and a body cavity. Mollusks were much more advanced animals than the trilobites.

Among the most important Ordovician fossils ever found are some tiny scraps discovered in Colorado sandstone. They date from about the middle of the period. They are the earliest known remains of animals with bone and may represent the first animals with backbones, called vertebrates. These early vertebrates are the ancestors of all of today's mammals, birds, reptiles, amphibians, and fishes. Fishes without jaws appeared during the middle of this period.

SILURIAN PERIOD

The Silurian was named after the Silures, an ancient tribe in Wales. This geologic period spans some 30 million years and lasted from about 440 million years ago to 410 million years ago.

Most of the present-day land areas of the Northern Hemisphere continued to be covered by shallow seas during the Silurian. The trilobites, highly successful for 150 million years, began to die out in large numbers. Giant water scorpions called eurypterids and measuring up to 10 feet (3 m) were common.

The largest animals of the Silurian belonged to that group called arthropods. Arthropods make up 80 percent of all known animals. Almost all have hard outside skeletons, like today's grasshoppers, lobsters, and spiders. They also have jointed legs that enable them to crawl, burrow, or swim. Some were flying insects. All had bodies divided into segments.

The giant sea scorpions of the Silurian probably fed on small bottom-dwellers called ostracoderms, which were covered with bony plates and may be the oldest vertebrates. During the Silurian the older arthropods ruled over the new-

Silurian landscape

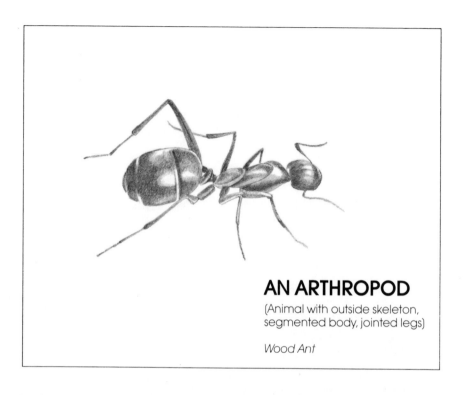

AN ARTHROPOD
(Animal with outside skeleton, segmented body, jointed legs)

Wood Ant

comer vertebrates. But that was not to last. Other organisms had developed thick, protective scales. Near the end of the Silurian new types of backboned animals were evolving. These were the placoderms and included fishes with jaws capable of biting, cutting, and crushing. This new feature—powerful jaws—made the placoderms especially successful hunters.

Notice that as we make our way up the geological time column, evolution produces organisms that are more and more complex in body parts. This increase in complexity of body plan also meant greater ability to move about and to respond to the environment in many other useful ways.

Something important to future life happened during the Silurian. Life took its first hold on the land. By this time, weathering of the surface rocks had formed soil, which allowed plants to take up life on the land. The earliest known fossils of land plants—ancestors of modern-day ferns—have been found in Silurian rocks in Australia and Europe. Called psilopsids, as a group they have survived up to present times. These leafless plants probably lived in shallow water and gradually evolved features that kept them alive as the level of the shallow inland seas lowered. Later, ferns were to evolve and become common land plants.

During the Silurian there was much volcanic activity in what is now Maine and the Canadian areas of New Brunswick and eastern Quebec. Land uplifting produced a 4,000-mile (6,500-km) -long mountain range that extended from Wales through Scandinavia and westward to northern Greenland. At this time it would have been possible to walk from Canada to Europe, since there was no Atlantic Ocean. Large deposits of salt, such as those found in western New York and Michigan today, were laid down at this time.

During the late Silurian, animal life seems to have taken to the land also. The earliest land organisms known are millipedes and scorpionlike animals such as those air-breathers found in late Silurian rocks from Scotland. Both types probably lived off the remains of sea animals stranded on the beach at high tide or during storms.

The Cambrian, Ordovician, and Silurian periods can be grouped together as the early Paleozoic era. It is very hard to reconstruct the details about the climate of 400 million years ago and spanning a period of about 175 million years. But fossils have provided us with some important clues. Since no one has produced any evidence of glaciers during the early Paleozoic era, we can guess that the climate was generally mild. Another reason for favoring a mild climate is the location of fossils of that time. Fossils of a given species are found

A fossil fern showing that our modern-day plants are descendants of an ancient species

widely spread both north and south. This would seem to tell us that there was hardly any difference in climate in what is now southern South America and mid-latitude Canada. Early Paleozoic fossils recently found north of the Arctic Circle differ very little from those found near the Equator. Again, the evidence points to a generally mild climate from the time Earth's first true cells arose until about 300 million years ago, which puts us well into the late Paleozoic.

CHAPTER FOUR

FROM FISHES TO AMPHIBIANS TO REPTILES

DEVONIAN PERIOD

The Devonian was named after Devon, England. This geologic period spans some 50 million years and lasted from about 410 million years ago to 360 million years ago.

The Age of Fishes is the name given to the Devonian, since these animals evolved into many forms at that time. Some were armored, like their Silurian ancestors. Others were covered with small scales. Some were sluggish and awkward, others streamlined, fast-swimming, and sharklike. The Devonian was a time of great change.

There were four main groups of Devonian fishes—two left over from Silurian times and two newcomers. The old Silurian types were the ostracoderms and the armored placoderms, some of which were terrors of the seas measuring 30 feet (9 m) long.

Of the two new types that had evolved, one was to become the ancestor of modern sharks and rays. These ani-

Devonian ocean landscape

mals did not have skeletons of bone but of a hard material called cartilage, which forms the "bone" of your nose. Some early members of this group had sharp teeth and were hunters.

The other of the two new groups was to give rise to the modern bony fishes of today's oceans, rivers, and lakes. Some of those early bony fishes had structures called ray-fins that were like the fins of a goldfish. They also had gills for breathing in the water.

A second group had an air bladder that worked as a lung for breathing air but had gills as well, for breathing in the water. These fishes also had powerful fins with a muscular lobe at the end of each fin. Within each lobe were bony structures like the bones of an arm or a leg. These lobes were strong and could be twisted and turned into many positions. They enabled the animals to move about on the land, which they most likely did when there was a need to go from a dried-up pond or streambed during a drought to a new body of water. This ability would be very important to the survival of those animals, since during Devonian times much of the land was a desert with little rain. So the first land animals were fishes in search of water!

During the wet periods the heavy rains eroded rock and washed it into rivers, which carried the sediments to the seas and formed great deltas. These sediments solidified as red sandstones. In deeper waters thick layers of mud were turned into slate and shale.

In addition to the fishes of the Devonian there were coral reefs built of limestone. Also there were many sponges, starfish, mollusks, and animals called sea lilies. Among the mollusks there were still many coiled cephalopods. By late Devonian times amphibians had evolved.

Amphibians are animals that spend part of their life cycle in water and part on the land. Traces or remains of them have been found in ancient swamps in Pennsylvania and

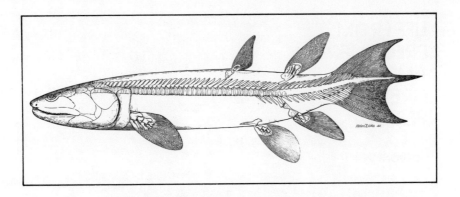

Lobe-finned fish used their fins to move over land between bodies of water. They represent an important link between fish and amphibians.

Greenland. They evolved from the fishes with lobe-fins, which became sturdy legs to move on.

By this time many other land organisms had evolved, including millipedes, scorpions, spiders, and the first insects, which were wingless. The warm climate of the Devonian allowed the slender plants of the Silurian to fill out into lush growths. Among other plantlife were horsetail rushes, tall tree ferns with stems more than 3 feet (1 m) thick, and forests of scale trees reaching heights of 45 feet (15 m). The Devonian brought Earth its first forests.

The trilobites and graptolites of earlier times began to die out during the Devonian. Another change marking this period was that high mountains were raised in what is now New England, Quebec, and Nova Scotia, as well as along the east coast of Australia. By the end of the Devonian and the beginning of the next period, the earlier supercontinent of Pangaea had broken into a northern landmass called Laurasia and a southern landmass called Gondwana. On both of

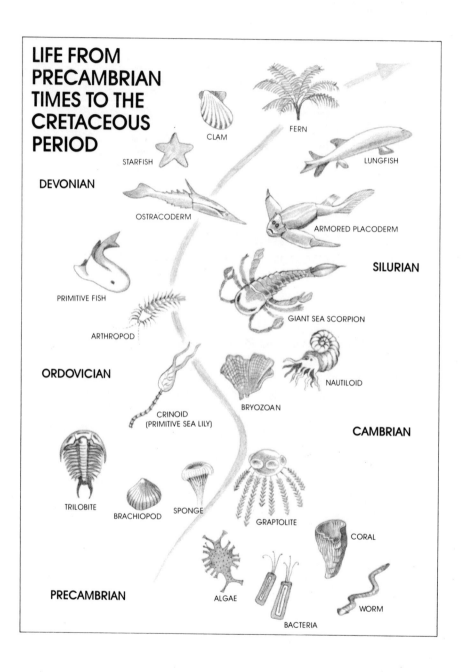

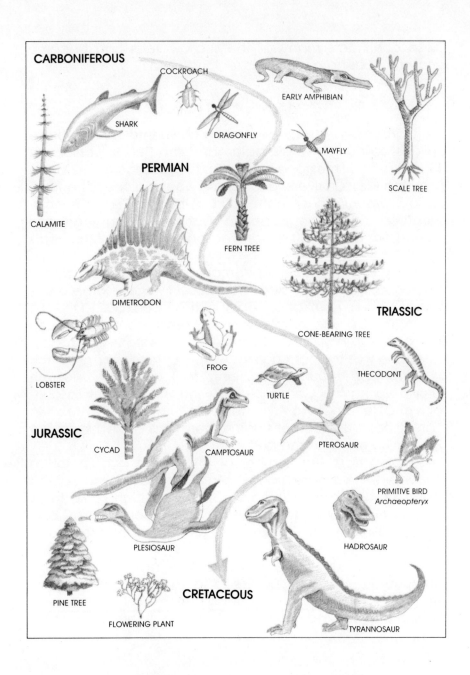

these continents many of the shallow seas had become smaller, exposing a much larger land area.

CARBONIFEROUS PERIOD

The Carboniferous may be broken down into two separate periods called the Mississippian and the Pennsylvanian, so named for the limestone area near the junction of the Mississippi and Missouri rivers, and for the coal regions of Pennsylvania. *Carboniferous* means "carbon-bearing"; coal is mostly carbon. This geologic period spans 75 million years and lasted from about 360 million years ago to 285 million years ago.

The land was changed in many important ways during this period, but shallow seas continued to cover large areas of the Northern Hemisphere early in the period. During the first part of the Carboniferous the shallow seas contained much lime and left limy muds that are the limestones of today. Limestone was the common rock type in the early part of this period. Much mountain building took place in western Europe during the Carboniferous, and the Ouachita Mountains of Oklahoma and Arkansas were formed and re-formed many times. The Southern Hemisphere was locked in an ice age during much of the period.

Judging from the number of fossil teeth and fin spines recovered, there were many sharks in the warm waters of the early Carboniferous. Hundred-foot (30-m) -high scale trees were common along the edges of pools, shallow lakes, and swamps. Below them was a dense undergrowth of ferns and other plants.

During the late Carboniferous some areas of the land sank and then rose, giving rise to large lakes and swamps. Gradually layer upon layer of dead trees and plants became matted and packed down in the swamps. It formed a mate-

An artist's depiction of a Carboniferous forest

rial called peat. Peat bogs are common in many parts of the world today and dried peat is sometimes used as fuel.

The formation of peat is the first stage in making coal. Time and again the land sank and then rose above the sea. Each time old forests died, new ones took their place. With each lowering of the land, seawaters flowed over the nearby peat-filled swamps, covering them with new sediment deposits that collected fossil remains. In time, chemical changes in the peat turned it into a soft brown coal; then, after being under pressure for millions of years more, the brown coal was changed into the coal we burn. Today we find layers of black coal alternating with layers of other sediments. About half of the world's usable coal was formed during the second half of the Carboniferous, mainly from the giant scale trees.

Life abounded during the Carboniferous. There were land snails and hundreds of species of cockroaches, some 4 inches (10 cm) long. There also were spiderlike animals such as centipedes and scorpions. Winged insects appeared for the first time; one species of dragonfly had a wingspan of about 30 inches (75 cm). Splashing in the swampy pools were lungfish and king crabs, whose descendants now live only in the seas.

The most interesting inhabitants of these Carboniferous pools were amphibians. Some were as small as a salamander. Others looked like snakes, and still others were the size of crocodiles. One biologist described the scene this way: "We can picture these sprawling creatures slithering among the fallen trees as they crossed from pool to pool, catching lung-

Many species of amphibians evolved in late Carboniferous.

fish and little enamel scales, snapping at great insects, nosing out a giant cockroach from under a log, and crunching up a king crab."

So many amphibians evolved during the late Carboniferous that this period is known as the Age of Amphibians. Amphibians were the only ruling animals with backbones until the end of the period, by which time one of their group had evolved into reptiles. Reptiles represent an important step in evolution. Amphibians reproduce by laying their eggs in water. Frog eggs, for example, develop into pollywogs, which can survive only in water. This is so because they breathe with gills, not lungs. Later, pollywogs develop lungs and legs, at which time they can move about on the land and breathe air. But to reproduce, adult frogs must return to water to lay their eggs. This is true of all amphibians.

When reptiles evolved, they had an important advantage over amphibians. A reptile's egg contains its own water supply within a watertight shell, so a baby reptile does not need a pond or puddle for its development. This meant that reptiles could occupy dry areas of the land, where amphibians could not live and where they would not be in competition with amphibians for living space and for food. This tipped the scales in favor of reptiles during the next period, when life away from water took many new turns.

PERMIAN PERIOD

The Permian is named after the province of Perm near the Ural Mountains of Russia. This geological period spans about 40 million years and lasted from about 285 million years ago to 245 million years ago. It marks the end of the Paleozoic era.

Like the Carboniferous, the Permian was also a time of great change. The Appalachians south of New England were thrust up, and the Ural Mountains of Russia were formed.

Permian landscape

Along the west coast of North America volcanoes poured lava out over the land. The reddish sedimentary layers of Monument Valley, Utah, were laid down during this time. Although what is now the western United States was still covered by shallow seas during this period, in other parts of the Northern Hemisphere inland seas were drying up. As they did they became more salty, making life in those seas even more difficult or impossible. As the seas eventually dried up, they left large deposits of salt and potash.

Most of the Permian forests were made up of cone-bearing trees similar to our present-day pines, firs, and spruce. The giant horsetails and scale trees of earlier times became smaller, although fern plants, tree ferns, and seed ferns continued to thrive.

Evidence for a general drying of the land in this period is suggested by the huge success of reptiles and a gradual dying out of amphibians. Without the widespread wet conditions of earlier times, amphibians could not thrive since they needed water in which to lay their eggs. This was not so of reptiles, who could safely lay their eggs on dry land. So a significant climate change favored the reptiles over amphibians.

There were many different kinds of reptiles. Some were small, quick, and lizardlike, while others were great lumbering beasts some 9 feet (3 m) long. Some ate plants, others ate insects, and still others ate other reptiles.

One called *Dimetrodon* was a fin-backed animal that once roamed the lowlands of Texas. The spines of its backbone were long and covered with a web of skin. Perhaps this "sail" acted as a radiator during the heat of the day and kept the animal cool. Then in the morning, after the animal had cooled during the night, the spread sail collected heat from the Sun and warmed *Dimetrodon* for its morning rounds. All reptiles, like amphibians, are "cold-blooded," meaning

Dimetrodon *was a fin-backed reptile that may have used its spiny backbone as a "sail" to keep cool in the heat of the day.*

Scratches in bedrock are the telltale signs of an ancient glacier.

that their body temperature is not self-regulating. Instead, it goes up and down with the air temperature.

The late Permian was a time of much dying among organisms living in the shallow seas. Over a period of only a few million years, at least half of these sea organisms became extinct. Gone were the trilobites, the ancient corals, most brachiopods, crinoids, bryozoans, and the placoderms, which had been the first true fishes.

What caused this mass extinction, the greatest known such event in the history of life on Earth? For a long time the answer remained a major biological mystery. But in the past years geological evidence seems to have solved the mystery. In the late Permian all the land in existence formed the single supercontinent of Pangaea. This caused the shallow seas that were around the edges of the earlier continents to be pinched off and dry up, or to become smaller. This could account for the extinction of species that had made those seas their home for millions of years. Also, it now seems that huge, broad ditches in the ocean floor caused a general lowering of the ocean level. Possibly during the late Permian the sea level dropped enough to expose the shallow seafloor shelves along the edges of Pangaea. Many species that had once thrived in the warm shallow water of those shelves would have been eliminated.

A telltale mixture of sand, pebbles, and clay shows that the Southern Hemisphere was gripped in an ice age for about 50 million years. It seems to have begun in the second half of the Carboniferous and probably lasted midway through the Permian, until about 260 million years ago. The unsorted sediments, called tillite, are made of material picked up by glaciers as they creep along and are dropped when the glaciers melt. Other telltale signs of a glacier are scratches in the bedrock.

The interesting thing about the late Paleozoic ice age is that those parts of South America showing evidence of the ice are now in tropical and semitropical regions. Geologists tell us that this is important evidence that the continents indeed have wandered about in the past just as they are wandering today.

Ice sheets covered not only South America but also South and Central Africa, India, Australia, and Antarctica. If those landmasses were once joined as Gondwana, the great southern continent that had drifted southward, then we could reasonably expect it to be covered with ice. It now seems that large-scale glaciation can take place only when a large continent remains in a polar position for a long time. This seems a more reasonable explanation than supposing that ice sheets can form at tropical and semitropical latitudes, where large temperature changes would have to take place, changes for which we have no evidence.

The Paleozoic era spanned some 400 million years and produced a dazzling variety of plants and animals. Its passing marked the end of the Age of Ancient Life and ushered in the Age of Middle Life. The two giant leaps in progress of animal life during the Paleozoic were the evolution of amphibians from lobe-finned fish, and the evolution of reptiles from amphibians. The reptiles were to evolve into those splendid creatures, the dinosaurs, who were to rule the land for more than 100 million years.

CHAPTER FIVE

LIFE DURING THE TRIASSIC PERIOD

The Mesozoic era, meaning the "time of middle life," spanned some 160 million years and is broken down into three periods—the Triassic, Jurassic, and Cretaceous.

The Mesozoic was a time of many important changes in life, environments, and geography. Most of the reptiles that had evolved earlier during the Paleozoic era were replaced by new forms that were very successful. Unlike their amphibian ancestors, these new reptiles did not need a sea, lake, or pond in which to lay their eggs. The eggs of reptiles could be laid on dry desert far away from water. This enabled the reptiles of the Mesozoic to spread far and wide, inhabiting every nook and cranny of the environment.

Some of these newcomers were more fantastic in appearance than the liveliest mind could imagine. In less than 100 million years after their first appearance, the reptiles came to rule the land and so wrote a remarkable evolutionary success story. They included the dinosaurs and fittingly made this era known as the Age of Reptiles.

Fossilized dinosaur eggs. Unlike amphibians, reptiles can lay their eggs far from water, on dry land.

CHANGES ON THE LAND

The Triassic period gets its name from the Latin word *trias*, meaning "three" and refers to a three-fold division of rock in southern Germany. This geologic period spanned about 35 million years. It began some 245 million years ago and ended about 210 million years ago.

It was during this period that the supercontinent Pangaea broke apart into a northern half (Laurasia) and a southern half (Gondwana). Seafloor spreading caused the ocean basins to greatly change.

All during the Triassic, plants and animals that produced calcium carbonate shells built up large deposits that later

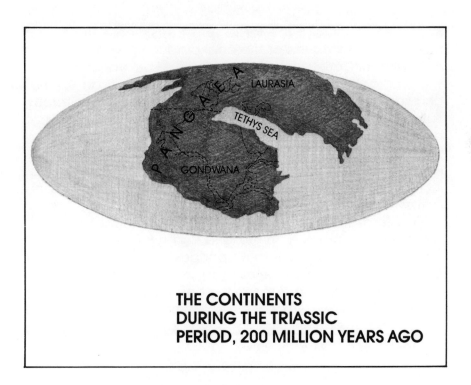

THE CONTINENTS DURING THE TRIASSIC PERIOD, 200 MILLION YEARS AGO

were turned into limestone. Chief among those organisms were coral and algae. The impressive Dolomite Alps of northern Italy contain limestone deposits laid down during the Triassic.

Early during this period the Appalachians were young mountains with impressive peaks, but, by the close of the Mesozoic, erosion had worn and rounded them into the rolling hills we know today. Also early during the Triassic, shallow seas covered the United States from the Pacific to central Utah and Wyoming. But a change in land level and a general drying was to soon change that.

CHANGES IN PLANTS AND ANIMALS

By the early Triassic, the large fern trees, scale trees, and most other vegetation of the coal-producing swamp forests of the earlier Carboniferous period had all but vanished. They were replaced by cone-bearing trees similar to today's pines and firs. These trees had evolved in the late Carboniferous. But a climate change from warm and moist conditions to cool and dry conditions favored the cone-bearing trees, which spread throughout the world in the Triassic. The famous Petrified Forest in Arizona is made up mostly of early Mesozoic fossilized cone trees. Many have trunks 5 feet (1.5 m) across and about 100 feet (30 m) long.

By Mesozoic times animals without backbones (invertebrates) were common on the land and in lakes and streams. There were clams, snails, tiny sponges, and brachiopods of earlier times. All ate algae as food. In the seas the cephalopods were still thriving. Among the cephalopods were a type called ammonites, which were so common and developed so fast that their fossil remains have been used in mapping the borders of Mesozoic seas.

The Petrified Forest in Arizona is made up of fossilized cone trees from the Triassic period.

Ammonites give scientists a way of mapping the borders of ancient oceans.

Just before the Mesozoic the highly successful trilobites died out. They were replaced by crustaceans such as shrimps, crabs, and lobsters, which appeared in the Triassic. There were also many starfish, sea urchins, and sea cucumbers in the early Triassic. All these creatures were much better at moving around than the trilobites. This gave them an advantage in catching food. You are more likely to find food if you go after it than by sitting still and waiting for it to pass you by, especially if there are a lot of animals after the same kind of food. Triassic sea animals that swam or crawled about increased in number. Those that lived attached to the seafloor became fewer.

Another evolutionary advantage that certain Triassic newcomers had over earlier forms were sharp teeth for cutting and tearing flesh. For example, the powerful arms and claws of the newcomer lobsters and crabs were an advantage over the much weaker limbs of the trilobites and king crabs that were common in earlier times. In the course of evolution, those animals that are better at competing for food generally do better than those less able to compete. Of two tribes of forest people, the tribe that hunts with bows and arrows has a better chance of surviving than the tribe that uses only its bare hands as a weapon.

During the Triassic there were many bony fishes living in both fresh water and in the oceans. But they were not as advanced as the fishes we know today. One major group of the time had diamond-shaped scales, scaly fins, and a lot of cartilage as part of their skeletons. As you will find, these fishes were to evolve in important ways.

Flying fishes first appear in this period. There were relatively few sharks during the Triassic, since many of their species had died out during the mass extinction late in the Paleozoic era. Without the widespread wetlands so common during the late Paleozoic, amphibians did poorly during the Triassic, especially the amphibian giants such as the flat-headed stegocephalians. Their fossil remains have been found clustered in dried-up lakebeds. Their name means "roof-headed." To open and close its mouth, a stegocephalian raised and lowered its upper jaw and skull while its lower jaw remained on the ground or lake bed with the rest of its bulky body.

Among smaller amphibians, modern frogs first appeared in the Triassic. By the close of the Triassic, most of the modern groups of insects had appeared.

SUCCESS OF THE REPTILES

The true success story that began during the Triassic was written by the reptiles. There were 75 percent more of their species during the Mesozoic than now, and they were to rule the land for 125 million years. So far, humans have ruled the land for only a few thousand years.

Many of the reptiles of the late Paleozoic did not make it into the Triassic and were replaced by hardier types. Turtles in their modern-day house of bone evolved during the Triassic. So did the ocean-living plesiosaurs and ichthyosaurs as well as the first crocodiles. Most of the reptiles that took up life in the sea had to return to the land to lay their eggs, as turtles do today. Ichthyosaurs (meaning "fish reptiles") were an exception. The eggs were not laid but hatched inside the mother's body.

Many new types of reptiles roamed the land. Small lizard-like animals called thecodonts ran about on their hind legs and lived mostly on a diet of grubs, snails, and insects. They were about the size of a collie dog and were an outstanding evolutionary success. Part of their success came from their way of life. Unlike most of the early reptiles, which were slow moving, the thecodonts were swift. Their large eyes provided excellent vision for spotting prey, which was always in plentiful supply.

Thecodonts were small, lizardlike reptiles that became the ancestors of the mighty dinosaurs that flourished in the Jurassic and Cretaceous eras.

The true giants among the reptiles, which were to shake the ground as they thundered overland during the Jurassic and Cretaceous periods, evolved from the thecodonts. And later, so did the flying reptiles, known as pterosaurs, and true birds.

It was during the middle Triassic that the first dinosaurs evolved. They were thin, two-legged members of a group called Saurischia. One of the earliest fossil finds of this animal was made in New Mexico. The hip region of saurischians was shaped like the hips of lizards. Some ate vegetation while others were meat-eaters and lived on dry uplands. The saurischians gave rise to some of the true giant dinosaurs of the later Mesozoic, including *Allosaurus* and *Tyrannosaurus rex*.

The second group of Triassic dinosaurs that was also to play an important role in the evolution of the Jurassic and Cretaceous giants was one called Ornithischia. The oldest remains of this group were found in Cape Province, South Africa. Their hips were shaped like the hips of birds. Most were plant-eaters that lived in marshes and lagoons. Later they gave rise to the water-loving dinosaurs known as the duckbills, which were plentiful during the Cretaceous. Although many of the saurischians had only tiny forelimbs and walked upright on two powerful hind legs, practically all of the ornithischians kept their powerful forelegs, as later represented by *Stegosaurus* of the Jurassic and *Triceratops* of the Cretaceous.

Around the mid-1900s more than 2,000 dinosaur tracks made by Triassic beasts were discovered on a construction

From dinosaur tracks such as those shown here, scientists have concluded that many dinosaurs traveled in herds.

site near Rocky Hill, Connecticut. Since the site was on public land, it was quickly made a state park. Because of the way the many tracks were laid out, scientists concluded that many of the dinosaurs that left the tracks roamed about in herds.

    Let us now look at the ways of life and the environmental conditions that shaped the lives of these fascinating monsters.

CHAPTER SIX

LIFE DURING THE JURASSIC PERIOD

The Jurassic period gets its name from the Jura Mountains of France and Switzerland. The mountains are made of rock of this period. The Jurassic spanned about 65 million years. It began some 210 million years ago and ended 145 million years ago.

CHANGES ON THE LAND

The Jurassic period was one of relative geologic quiet, although the Sierra Nevada Mountains were formed then. Also, the western edges of both North and South America sank. The climate seems to have been one of warm and moist conditions favoring many swamps and forests in contrast to the drier climate of the Triassic. In a way, the Jurassic was a stage-setting period for the great activity that was to come during the Cretaceous.

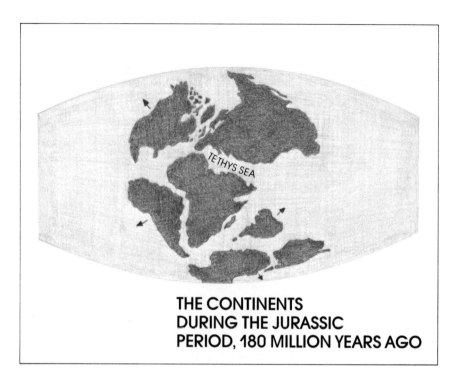

THE CONTINENTS DURING THE JURASSIC PERIOD, 180 MILLION YEARS AGO

By the beginning of the Jurassic, North and South America had moved apart by several hundred miles. At the time, North America was still attached to Africa. But as North America pulled away about 190 million years ago, a piece of Africa stuck to it and became Florida. Florida's oldest rocks and fossils are much more like those of Africa than those of the rest of the United States. Later in the period North America slid westward as seafloor spreading widened the Atlantic Ocean bed. Meanwhile, both Africa and India split apart from Antarctica. By the end of the Jurassic, South America and Africa had begun to split apart as separate continents.

During most of the Mesozoic the South Pole was probably located in the southern Pacific Ocean and the North Pole was near present-day Japan. As long as the poles were in those positions, glacial ice could not form, so the climate during the Mesozoic was generally milder than our climate today.

One of the major long-term events during the Mesozoic was the steady deposit of sediments all along an enormous geosyncline trench called the Tethys Seaway. The trench stretched for thousands of miles between the southern edge of Europe eastward to Asia and the northern edge of Africa and India. During the era following the Mesozoic the sediments dumped during the Triassic, Jurassic, and Cretaceous into the Tethys Seaway were to be thrust up as the Alps, Pyrenees, Apennines, Atlas, Carpathians, Caucasus, Pamirs, and Himalayan mountain ranges.

CHANGES IN PLANTS AND ANIMALS

Botanists (plant scientists) call the Jurassic the Age of Cycads. Cycads are palmlike trees with a great cone at the top. They provided a major food source for many of the plant-eating dinosaurs. Ferns and plants called scouring rushes, or horsetails, were also common during this period.

Late in the Jurassic the first pine trees appeared, as did giant sequoias. When we gaze on these giant marvels of the plant world in the redwood forests of California, we should remember that, as a group, they have survived for some 200 million years.

Ammonites, which had appeared in large numbers during the Triassic, became even more widespread during the Jurassic. There were so many of them, and so many different kinds, that scientists have been able to link certain rock

zones with certain kinds of ammonites. So ammonites give us a way of mapping rock layers. Geologists in England list about 50 ammonite zones. Ten such zones have been found in the United States, and 24 in Mexico. A period of about one million years separates one zone from another.

Organisms called protozoans also evolved in large numbers during the Triassic and became widespread during the Jurassic. Protozoans are "animals" consisting of a single cell that carries out all life functions. They include amoebas and paramecians. Some types of amoeba developed a protective shell of calcium peppered with tiny holes. Called foraminiferans (meaning "bearers of windows"), some of these organisms poke sticky, threadlike tentacles out through the tiny holes and catch passing bits of food. Others, called radiolarans, had protective shells of sand. There were at least twenty-five major groups of protozoans during the Jurassic.

Among the fishes, the heavy-scaled, soft-boned forms that had evolved during the Triassic were replaced by new forms. The newcomers of the Jurassic had thinner scales, evenly shaped tails, and the hardest bone in their skeletons. These fishes were the ancestral parents of our present-day fishes. Sharks, which had largely died out at the end of the Paleozoic, made a dramatic comeback during the Jurassic and once again became terrors of the seas. Rays were also plentiful.

Jurassic algae of the sea produced huge amounts of small lime pellets. Stuck together, they formed rocks called

Giant sequoias first appeared in the late Jurassic and have survived as a group for more than 200 million years.

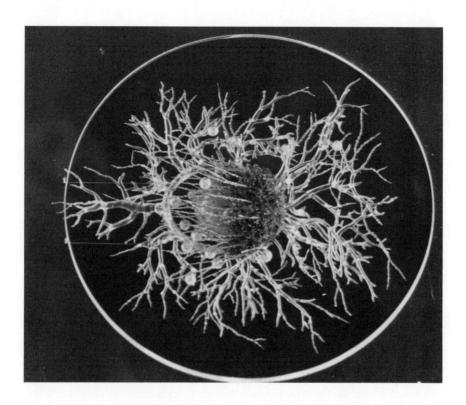

A glass-blown model of a protozoan

oolite. The bedrock of Florida is made up largely of oolite. The seas also had many coral reefs and other reefs formed by sponges. Mollusks, such as clams and oysters, as well as the cephalopods were common.

The rulers of the Jurassic seas were not giant fishes but air-breathing reptiles that had taken to the water. Among these were the plesiosaurs, great monsters up to 30 feet (10 m) long who paddled the shallow water along the shores in search of a meal of fish or cuttlefish. Most of the Jurassic plesiosaurs had small heads, long necks, and arms like those of a turtle. The ichthyosaurs of the Triassic seas evolved into more streamlined animals during Jurassic times. They drove themselves through the water at high speed with tremendously powerful tail fins. Other reptiles that prowled the Jurassic seas included crocodiles, with long snouts in search of fishes, and many kinds of turtles.

Just as the Triassic provided a preview of the magnificent reptiles to evolve during the Jurassic, the Jurassic provided a preview of the splendid group of dinosaur giants to evolve during the Cretaceous. Lizards evolved during the Jurassic, but snakes were not to appear until later. Some of the Jurassic dinosaurs were armored, such as one kind named *Scelidosaurus*, which had the appearance of an army tank. *Scelidosaurus* was a plant-eater that lived on the edge of swamps. Another dinosaur, called *Camptosaurus,* weighed several tons and lumbered along on muscular hind legs.

The giants of the Jurassic were all plant-eaters. These included *Diplodocus, Brachiosaurus, Cetiosaurus,* and *Apatosaurus,* all weighing 20 or more tons and measuring 60 feet (18 m) long. Swamps were the homes of these giants that sloshed along on four powerful legs like tree trunks. They had a special nerve center in their hips that controlled their walking. This nerve center was bigger than the brain in their small heads. The swamps were not only a source of food for these animals but protected them from flesh-eating land-dwellers such as *Megalosaurus* and *Antrodemus,* which had cruel hind claws for killing their victims. Although smaller than the swamp-dwellers, *Megalosaur* and *Antrodemus* moved about swiftly on their hind legs and had sharp teeth with which to tear apart flesh.

REPTILES THAT FLEW

The Jurassic saw the first flying reptiles, the pterosaurs. Somewhat like bats, they had wings formed by long arms and

Plesiosaurs (above), who were rulers of the Jurassic seas, and the swamp-dwelling Apatosaurus (Brontosaurus)

clawlike fingers attached to a sheet of skin. They flew and glided about in search of insects, occasionally diving to snatch a fish out of the water. They could walk as well as fly, but like an airplane they needed a takeoff run to become airborne.

The earliest known bird, called *Archaeopteryx,* evolved during the late Jurassic. Fossil remains of two were found in Germany. About the size of a crow, *Archaeopteryx* had true feathers, but its body plan was that of a reptile, including a reptilian jaw with teeth and a reptilian tail. Some students of evolution call *Archaeopteryx* a "flying dinosaur."

Scientists search for such fossils that show how one group (in this case, reptiles) evolves into a different group (birds). There are enough such examples of so-called transition types to prove that evolution of one type of organism into another type has taken place. So far in our account of the pageant of life through geologic time, we have seen that fishes evolved into amphibians, that amphibians evolved into reptiles, and that reptiles evolved into birds.

Although birds became a marvelous evolutionary success, pterosaurs did not. Pterosaurs were not fliers but gliders. When damaged, the skin forming their wings may have been hard to repair, although today's bats, which also have wings of skin, don't seem to have a problem. Birds, on the other hand, were true fliers, and their feathered wings were easily repaired by growing new feathers individually to replace damaged ones.

Perhaps the most important thing to happen in the late Triassic and early Jurassic was the appearance of the first mammals. Mammals were a new type of animal, one that had body hair, gave birth to their young rather than laying eggs, provided milk for their young, and kept a steady body temperature.

Like birds, mammals also evolved from reptiles. Humans, dogs, cows, whales, and rabbits are among the many kinds

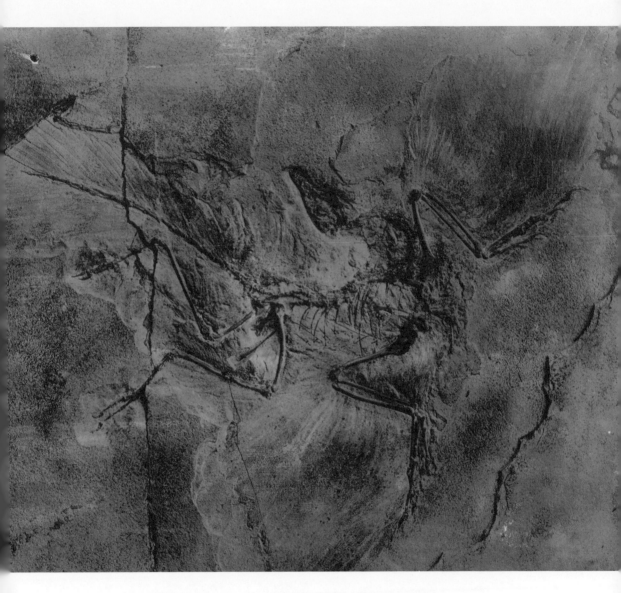

A fossil imprint of Archaeopteryx, *the earliest known bird, which evolved during the late Jurassic.*

of mammals. Fossils of the earliest known mammals are scarce. Scientists have found only jaws and bits and pieces of other bones in South Wales, East Africa, and China.

The mammals of the Mesozoic were about the size of a mouse or a rat. They lived in the undergrowth or the branches of trees and had to rely on their swiftness to avoid being eaten by the many flesh-eating reptiles.

Among the more exciting fossil finds of mammal-like reptiles was one made in northwestern Nova Scotia in 1985. The discovery turned up more than 100,000 pieces of skulls, teeth, jaws, and other bones of dinosaurs, ancient crocodiles, lizards, sharks, and early fishes. Also found were tiny dinosaur footprints the size of a penny—belonging to the smallest known dinosaurs. But the most significant fossils were 12 skulls and jaws of mammal-like reptiles called tritheledonts. The animals were about a foot long and are the first such fossils found in North America. They may help provide an answer to where the first mammals came from—tritheledont, or from an earlier common ancestor of both mammals and reptiles.

As the dinosaurs exploded into a colorful display of different types during the Mesozoic, the mammals were to do the same after the close of the Mesozoic and become rulers of the land. But before that happened the dinosaurs were to reach their peak in the Cretaceous.

CHAPTER SEVEN

LIFE DURING THE CRETACEOUS PERIOD

The Cretaceous period gets its name from the Latin word *creta,* meaning "chalk," since enormous chalk deposits were laid down in Europe and North America during this time. The Cretaceous spanned some 80 million years. It began 145 million years ago and closed 65 million years ago.

CHANGES ON THE LAND

By the end of the Cretaceous Earth's geography had changed greatly from earlier Jurassic times. Seafloor spreading had split South America and Africa by a distance of some 1,700 miles (2,800 km). Europe and North America also had split apart northward, leaving a small plug of land that was to become Greenland. The northward movement of Africa had narrowed the Tethys trench, which continued to collect sediments throughout the Cretaceous. India was moving northward on a course that would eventually join it to Asia.

Antarctica had moved southward to the South Pole but had not yet separated from Australia. The pattern of continental drift and seafloor spreading still taking place today was set in motion during the Mesozoic.

Cretaceous seas covered most of Europe, much of Asia, and nearly half of North America. This flooding was caused by a sinking of the land. The Gulf of Mexico received nearly 13,000 feet (4,000 m) of sediments during this period. The Rocky Mountains of North America and the Andes of South America were thrust up out of geosyncline trenches that had been filling up with sediments for millions of years.

If you happen to live anywhere from northern New Jersey to Maryland, the rock exposed at the surface is from the early Cretaceous. Rocks of the Cretaceous also rim the Pacific Ocean—from the islands of Japan, south to Korea and Formosa, eastward to the East Indies and New Zealand, then northeast to the Andes of South America and the Rockies of the United States.

The Pacific Ocean floor also contains many stumps of volcanic mountains whose tablelike tops were cut off by ancient erosion and that are from the Cretaceous age. These stumps are called guyots, or tablemounts. Fossils of Cretaceous land organisms have been found at the tops of many of these submerged volcanic mountains, proving that in earlier times their tops were above water. So about halfway through the Cretaceous the rim of the Pacific Ocean basin must have been alive with volcanoes and earthquakes, hence its name "the rim of fire."

CHANGES IN PLANTS AND ANIMALS

The plants of the Cretaceous were not very much like those of the Jurassic. Nearly gone were the cycad forests of old, replaced by rapidly spreading pine forests. But the most important evolutionary event among plants was the appearance of plants that produced flowers. Among them were many new trees—elms, oaks, maples, and others we know today.

Flowering plants had a certain advantage over the older trees with cones. The seeds produced by cones are exposed on the surface of the cone scales. This means that they can be easily destroyed. The seeds of the flowering plants were safely enclosed within a special compartment of the flower. This advantage is one reason for the very rapid spread of flowering plants around the world during the Cretaceous. Forests of these newcomers became so widespread that they gave rise to Cretaceous coalbeds. The land was beginning to resemble today's tropics, and we might almost have felt at home then.

By the time of flowering plants, there were many kinds of insects, and interesting relationships between certain insects and flowering plants evolved that gave a further evolutionary advantage to the flowering plants. Bees, moths, wasps, beetles, flies, birds, and even bats are attracted to flowers to gather nectar, a sugar substance that bees turn into honey. Whenever a bee visits a flower, pollen grains stick to the bee's body. Then, when the bee visits a different flower, it rubs some of the pollen onto that flower. In that way bees and other insects carry pollen from flower to flower. Pollen is a substance flowering plants need to produce more plants. If a flowering plant pollinates itself, its offspring will be just like the parent plant. But when a flowering plant is pollinated by a different plant, its offspring can end up with differently shaped or colored flowers, shorter or longer stems, and so on. Cross-pollination, instead of self-pollination, produces a hardier species of plant.

The oldest known ants evolved during the late Cretaceous. The body of one was found perfectly preserved embedded in a hardened drop of tree pitch, called amber.

Globigerina, *a protozoan from the Cretaceous, is still around today.*

Species of mollusks, which had spread during the Jurassic, became more numerous and varied during the Cretaceous. There also were many air-breathing snails, lots of freshwater clams, and large reefs made up of oysters. The ammonites, so successful during the Jurassic and early Cretaceous, died out at the end of this period. Protozoans reached spectacular numbers. A newcomer, called *Globigerina*, floated in the open seas, as it continues to do today. The soft, chalky shells of *Globigerina* coat the seafloor.

Cretaceous seas were alive with bony fishes that had gradually evolved thinner scales, more efficient tails and fins, and harder skeletons. These are the same types of fishes that swim in today's oceans and rivers. Sharks continued from the Jurassic and prowled Cretaceous seas in search of prey. Although ichthyosaurs were still around, there were not as many as during Jurassic times.

The Cretaceous was the heyday of the plesiosaurs. One group of these sea monsters, called elasmosaurs, had necks about 20 feet (6 m) long. They were too bulky to be swift swimmers, but their whiplash necks could strike prey with the swiftness and accuracy of a snake.

Giant crocodiles continued to swim the Cretaceous seas and were joined by a newcomer, marine lizards called mosasaurs. These monsters resembled the mythical sea serpents. They reached 50 feet (15 m) in length and had snake-like jaws that could unlock and swallow large prey. Their fossils have been found the world over, but they were short-lived, appearing only during the late Cretaceous period.

Over: *Mosasaurs joined other marine lizards in the Cretaceous seas. In the air above are* Pteranodons, *the flying reptiles.*

The last major group of reptiles to evolve were the snakes, which can be regarded as lizards that lost their legs. Fossils of snakes have not been found earlier than Cretaceous times.

Pterosaurs continued into the Cretaceous and became larger. Some, such as *Pterodactylus*, were as small as sparrows. At the other extreme was the *Pteranodon*, with a wingspan of 25 feet (8 m) and weighing about 60 pounds (30 kg). Like birds, it had hollow, air-filled bones. It was a glider rather than a flier.

Before the close of the Cretaceous a dozen or more species of birds had evolved. Unlike the flying reptiles, birds had powerful breast muscles designed for repeated flapping of the wings for true flight instead of only gliding. The pterosaurs lacked such muscles. Among Cretaceous birds were swimming birds, one nearly 5 feet (1.5 m) long. Fossil ancestors of flamingos and geese have been found in Cretaceous rocks.

LAND GIANTS OF THE CRETACEOUS

The tyrant of the land was the flesh-eating monster, *Tyrannosaurus rex*. From its nose to the top of its powerful tail it measured up to 50 feet (15 m). Its 4-foot (1.2-m) -long head had powerful jaws with many sharp teeth designed for tearing flesh. It moved about on tremendously powerful hind legs and had forelegs that were too short to be used for running.

Among the chief prey of *Tyrannosaurus* were the duck-billed dinosaurs called hadrosaurs, who lived close to rivers and lakes and used their shovel-shaped bills to scoop up plants growing on the bottom. Hadrosaurs had toes with hoofs and webs of skin between their toes. Some had bony

Tyrannosaurus rex

Triceratops

helmets on the top of their heads. Another true giant that walked along upright was the iguanodon, a plant-eater that spent most of its time in the forests munching on leaves and small branches.

Although the duck-bills, iguanodon, and *Tyrannosaurus* all had tough, leathery skin, most of the other land dinosaurs had bony plates of armor and spikes for protection. *Triceratops* was perhaps the most splendid of these armored monsters. It was about 30 feet (10 m) in length and plodded along on four thick legs. It had a great collar of bone with a large horn over each eye and another horn on the end of its nose.

Like the reptiles of today, all of these Cretaceous beasts were cold-blooded. This meant that they would be sluggish in the cool early morning hours and generally inactive until the Sun warmed them. Then, in the cool of the evening, they would once again become inactive. This was not so of the newcomer mammals, who were warm-blooded and could be active whenever it suited them to be.

This was to be a great evolutionary advantage that was to help tip the scales in favor of the mammals in the next geologic era. But it was the dinosaurs who continued to rule until the close of the Cretaceous, which also marked the close of the Mesozoic era.

And then they were gone. By the close of the Cretaceous the dinosaurs and flying reptiles had vanished, not a trace of them to be found in sedimentary rock anywhere younger than 65 million years. Gone also were the ocean-going plesiosaurs, mosasaurs, and ichthyosaurs, and along with them the ammonites that had been so plentiful earlier. The only reptiles to survive this mass extinction were the crocodiles, lizards, turtles, snakes, and the strange tuatara of New Zealand. The great science mystery that has yet to be solved is why the dinosaurs became extinct.

THE EXTINCTION DEBATE

There have been many theories, but none by itself seems able to completely solve the mystery. Some have wondered if a large comet or asteroid might have struck Earth and caused the mass extinction. If so, then why did snakes, crocodiles, lizards, and other forms of life survive?

Others have wondered if the increasing number of small mammals could have drastically reduced the supply of dinosaur eggs. Or could continental drift and mountain building have caused a cooling of the climate that was too much for the dinosaurs? Fossil plant remains show some cooling, but not enough to have wiped out the dinosaurs.

Could widespread disease have been the cause? Each new theory seems only to confuse the picture more rather than offer an answer.

In the ongoing debate over mass extinctions, there are two major theories:

1. Mass extinctions may be caused by catastrophic events, such as a large asteroid or a large comet crashing into Earth. This event, it is argued, would produce several large-scale damaging results. Dust raised by the explosive impact would be carried high into the air and form a dust cloud enclosing the planet. This would cause dark, reddish skies and a worldwide cooling that could wipe out land species that are especially sensitive to temperature change, such as those living in a tropical climate. Another result of a large comet or asteroid impact, according to some scientists, would be extensive showers of acid rain that would kill many plants and animals. The acid rain, they say, would be caused by the sudden heating of the atmosphere, which would cause oxygen to combine with nitrogen and produce nitric acid, the harmful chemical in acid rain.

2. Mass extinctions may be graded rather than sudden, say other scientists. No matter what the cause—a comet

impact or a change in energy output of the Sun, for example—the environment is changed in major ways. According to this theory of graded extinctions, environmental change takes place gradually over a few million years. Those species that would disappear first would be ones adapted to very specific environments and so could not survive a drastic change; for instance, cold-blooded animals, or fishes very sensitive to water temperature, salinity, or acidity. Next to go would be those species adapted to a greater diversity of environments and that might survive for a while, but not over many centuries.

BENEFITS FROM THE MESOZOIC

Many geological events of the Mesozoic are important to our lives today. More oil was formed in Mesozoic rocks than in any other period. A little more than 50 percent of the world's known oil supplies is locked up in Mesozoic rocks. The rich oil fields of the Middle East are nearly all in Jurassic or Cretaceous rocks. Coal also has been found in Triassic, Jurassic, and Cretaceous rocks.

Cretaceous deposits of coal are especially rich in the United States and Canada. Sedimentary rocks (sandstone and shale) formed in the Triassic in Germany, Russia, and the United States contain copper deposits. Jurassic rocks in England and France contain iron ore. In the western United States, Triassic and Jurassic rocks hold rich stores of uranium.

The gold that caused a stampede of thousands of miners to California in the 1800s was formed during late Jurassic times. So were the rich salt domes and sulfur deposits of the Gulf Coast. More salt is found in early Cretaceous deposits in Brazil, for example. The famous diamond mines of South Africa were made possible by the formation of the diamond-containing rocks in Cretaceous times, but the diamonds themselves were probably formed earlier.

So the Mesozoic provided not only a wealth of new animal and plant species through evolution but also an impressive wealth of minerals. Many of the reptiles and amphibians familiar to us today, along with flowering plants, pine trees, and sequoias, appeared during the Mesozoic and have remained almost unchanged to this day.

GLOSSARY

Adaptation. The condition of a plant or animal population being in tune with its environment, or its ability to adjust to changes in the environment (scarcity of food, or change in climate, for example).

Amino acids. Complex molecules that were among the first living molecules. Amino acids contain carbon, oxygen, nitrogen, and hydrogen. These molecules are the building blocks of proteins. There are about 20 different kinds of amino acids (see *proteins*).

Amphibians. The animal group that spends part of its life cycle in water and part on land, including frogs and salamanders. Amphibians lay their eggs in water, where the eggs hatch and the young are fishlike and breathe through gills. Later the young develop into land-dwellers with lungs and four legs.

Arthropods. The animal group that includes 80 percent of all known animal species. Almost all arthropods have hard outside skeletons, have jointed legs that enable them to crawl, burrow, or swim, and have bodies divided into segments.

Botanist. Any scientist who specializes in the study of plants.

Cambrian period. The geologic period spanning 85 million years and lasting from about 590 million to 505 million years ago.

Carboniferous period. The geologic time period spanning 75 million years and lasting from about 360 million years ago to 285 million years ago.

Cell. The smallest organized unit of living matter recognized by biologists. All living organisms are composed of cells. Some organisms, such as a paramecium, are a single cell.

Cretaceous period. The geologic time period spanning some 80 million years and lasting from 145 million years ago to about 65 million years ago.

Devonian period. The geologic time period spanning some 50 million years and lasting from about 410 million years ago to 360 million years ago.

Dinosaur. Any of the many extinct reptiles that lived during the Mesozoic era, some of which reached gigantic size. For reasons still unknown, all of the dinosaurs had become extinct by about 65 million years ago, at the end of the Cretaceous period.

Diversity. The many different kinds of animal and plant species that have evolved over the past 3 billion or so years. Scientists have classified more than 1,200,000 different animal species and at least 500,000 species of plants. Each year thousands of newly discovered species are added to the lists.

Evolution. The process that explains the various patterns of biological change that ultimately causes the success (adaptation) or failure (extinction) of species and produces new species of plants and animals.

Extinction. The total disappearance of an entire species. Once a species has become extinct, it is gone forever.

Glucose. A sugar-food used by plants and animals alike. Glucose is produced by green plants when they combine carbon dioxide and water vapor in the presence of light as an energy source. In the process, the green plant gives off oxygen as a by-product to the air.

Guyot. The stump of an undersea volcanic mountain with a flat top; also called a tablemount.

Invertebrates. Any animal species lacking a backbone. The first animals to evolve were invertebrates. Early invertebrates that lived in the ancient seas included sponges, trilobites, brachiopods, and graptolites.

Jurassic period. The geologic time period spanning about 65 million years and lasting from about 210 million years ago to 145 million years ago.

Mammal. Any vertebrate animal that has warm blood, a covering of hair, gives birth to its young (with two exceptions), and suckles its young.

Membrane. A protective and porous "jacket" enclosing cells. A membrane separates an organism's inside environment from the outside environment. The development of a membrane was an important step in the evolution of the first biological cells some 3 billion years ago.

Mesozoic era. The time of "middle life," spanning some 160 million years and usually broken down into three periods—the Triassic, Jurassic, and Cretaceous. The dinosaurs reached their peak during the late Mesozoic.

Mutation. A random, or chance, change in a plant or animal's genes that make the organism different in one or more ways from its parents. Most mutations are harmful, but a few are beneficial. These changes may be passed on to offspring.

Oolite. A type of rock formed from little pellets of lime produced by Jurassic algae. The bedrock of Florida is made up largely of the rock Miami oolite.

Ordovician period. The geologic period spanning about 65 million years and lasting from about 505 million years ago to about 440 million years ago.

Paleontologist. A scientist who specializes in the recovery and study of fossils.

Paleozoic era. The period of "ancient life," predating the Mesozoic era and spanning 325 million years of Earth's history. The Paleozoic is made up of seven geologic periods.

Peat. A material formed in swamps, from dead trees and other vegetation that has been packed in layers over many years. Dried peat is sometimes used as fuel. Peat represents the first stage in the formation of coal.

Permian period. The geologic period spanning about 40 million years and lasting from about 285 million years ago to 245 million years ago.

Precambrian era. The geologic time span from 4.6 billion years ago—the estimated age of Earth—to 570 million years ago. The Precambrian is generally viewed as the time when life was firmly establishing itself on Earth.

Proteins. Substances made up of amino acids. Our bodies use proteins for growth, repair of injured parts, and as a source of energy.

Silurian period. The geologic time period spanning some 30 million years and lasting from 440 million years ago to 410 million years ago.

Reptiles. Cold-blooded vertebrates, including lizards, snakes, and alligators. The Cretaceous period marked the peak of the reptiles' success, as the dinosaurs, which were reptiles, thrived and dominated Earth.

Rift valley. A fracture in Earth's crust through which magma from the mantle rock layer oozes out onto the surface. The Mid-Atlantic Ridge, running north and south along the Atlantic Ocean, is a rift valley.

Species. Any one kind of animal or plant group, each member of which is like every other member in certain important ways. All populations of such a group must be capable of interbreeding and producing healthy offspring.

Stromatolites. Dome-shaped fossil remains made up of hundreds of wafer-thin layers. These little mounds, found in western Australia, for instance, are the fossil remains of colonies of primitive bacteria.

Tillite. An unsorted mixture of sand, pebbles, and clay that provides us with evidence of ice ages. Tillite consists of materials picked up by ancient glaciers.

Triassic period. The geologic time period spanning about 35 million years and lasting from some 245 million years ago to about 210 million years ago.

Variation. The racial and certain other differences among the individuals making up a population. These variations are what lead to evolutionary change.

Vertebrate. Any animal species having a backbone.

INDEX

*Italicized page numbers
indicate illustrations*

Adaptation, 17–18, 36, 57
Africa, 10, 12, 65, 75
Ages
    of amphibians, 44
    of ancient life, 50
    of cycads, 65
    of fishes, 9, 34
    of middle life, 50
    of reptiles, 51
Air-breathers, 31
Algae, 12, 54, 67
Allosaurus, 60
Ammonites, 54, *56*, 65, 67, 79
Amoebas. *See* Protozoans
Amphibians, 36–37, 43–44, 46, 50, 57–58
Antarctica. *See* South Pole
Antrodemus, 69
Ants, 77

Apatosaurus, 69
Archaeopteryx, 72, *73*
Arthropods, 28, *30*
Asia, 75, 76
Australia, 10, 12, 13, 31, 37, 76

Birds, 72, *73*, 82
Birth, 72; *see also* Egg laying
Botanists, 65
Brachiopods, 27, 49
Brachiosaurus, 69
Bryozoans, 27, 49

Cambrian Period, 9, 22–25
Camouflage, 17
Camptosaurus, 69
Carboniferous Period, 16, 40–44, 49
Cells, 14
Cephalopod, 27, 36, 54, 68
Cetiosaurus, 69
Chalk, 75
Chemicals, 10–14

(94)

Climate, 24, 31, 33, 37, 63, 65, 86
Coal, 40, 43, 77, 87
Cold-blooded, 46, 49, 85
Continental drift, 50, *64*, 75, 76
Cretaceous Period, 51, 75–88
Crinoids, 49
Crocodiles, 58, 69, 79
Cycads, 65

Darwin, Charles, *16*, 19
Devonian Period, 34–40
Dimetrodon, *46–47*
Dinosaurs, 19, 50, 60–*61*, 82, 85–87
Dinosaur tracks, 60–62
Diplodocus, 69
Diversity of life forms, 16
Duckbills, 60

Egg laying, 44, 46, 51, 52, 58
Elasmasaurs, 79
Eras. *See* Mesozoic Era; Paleozoic Era; Precambrian Era
Europe, 31, 40, 75, 76
Eurypterids, 28
Evolution, 18–20, 24, 30, 44, 50, 58, 69, 72
   defined, 19
   of mammals, 72
   of plants, 31, 76
   of reptiles, 82
   theory of, 19–20
Extinction, 16–19, 28, 49, 57, 85–87

Ferns, 31, *32*, 37, 46
Fish, 34–*37*, 43, 49, 50, 57, 67, 79; *see also* Placoderms
Flowers, 77
Flying fish, 57
Flying dinosaur. *See* Archaeopteryx
Flying reptiles. *See* Pterosaurs
Food, 27, 46, 54, 57, 60, 68, 69, 72
Forests, 37, 40, *41*, 46, *55*, 65, 77
Fossil record, 9, 20
Fossils
   Cambrian, 23–*25*

Cretaceous, 76
Ordovician, 28
Paleozoic, 31–32
Precolumbian, 9, 12
Frogs, 44, 58

Glacier, 48
Globigerina, 78, 79
Gondwana, 37, 40, 50, 53
Graptolites, 37
Greenland, 31, 75
Guyots, 76

Hadrosaurs, 82, 85
Horsetails, 65

Ice age, 40, 48, 49, 50
Ichthyosaurs, 58, 69, 79
Iguanodon, 85
India, 64, 75
Insects, 37, 43, 58
Invertebrates, 24, 54

Jurassic Period, 51, 60, 63–74

Lakes, 40
Land, 24, 31, 40, 43, 46, 76
Land organisms, 31, 37, 43; *see also* Amphibians; Mammals; Reptiles
Laurasia, 37, 40, 53
Limestone, 36, 40, 53
Lizards, 69, 79

Mammals, 72, 74
Mass extinctions, 49, 85–87
Megalosaurus, 69
Mesozoic Era, 38–39, 51–58
Minerals, 87–88; *see also* Coal; Oil
Mollusks, 27–28, 68, 79
Mosasaurs, 79, *80–81*
Mountains, 25, 40, 45, 54, 63, 65, 76, 86
Mutation, 17–18

North America, 12, 24, 25, 31, 37, 46, 63, 64, 67, 74–76

North Pole, 64

Oil, 87
Oolite, 68
Ordovician Period, *26*–28
Ornithischia, 60
Ostracoderms, 28, 34

Paleozoic Era, 21–32, 49, 58
Pangaea, 37, 49, *53*
Paramecians. *See* Protozoans
Peat, 40, 43
Permian Period, 44–50
Placoderms, 30, 34, 49
Plants, 37, 76–77; *see also* Ferns
Plesiosaurs, 58, 68–71, 79
Pollination, 77
Precambrian Era, 9–21, 24
Protozoans, 67, 68, 79
Pteranodon, 80–81, 82
Pterodactylus, 82
Pterosaurs, 60, 69, 72, 82

Redwood trees, 65
Reptiles, 45, 50, 51, 58, 68–69, 82
Rock formation, 67

Salt deposits, 31, 46, 87
Saurischia, 60
Scelidosaurus, 69
Sea level, 49, 63
Seafloor spreading, 53, 64, 75–76
Seas, 9, 24, 76, 79
Shale, 25, 36
Sharks, 34, 40, 57, 67, 79

Silurian Period, 28, *29*
Slate, 36
Snakes, 82
South America, 49, 63, 64, 75
South Pole, 64, 65, 76
Species, 16, 19
Stegocephalians, 58
Stegosaurus, 60
Stromatolites, 10
Supercontinent. *See* Pangaea
Survival of the fittest, 17–18, 19, 24, 36, 44, 46, 57, 77, 85

Tablemounts. *See* Guyots
Tethys Seaway, 65, 75
Theocodonts, 58, 59
Tillite, 49
Time periods. *See* Cambrian; Carboniferous; Cretaceous; Devonian; Jurassic; Ordovician; Permian; Silurian; Triassic
Trees. *See* Forests
Triassic Period, 51–*55*
Triceratops, 60, 84, 85
Trilobites, 24, 25, 27, 28, 49, 57
Tritheledonts, 74
Turtles, 58, 69
Tyrannosaurus rex, 60, 82, *83*, 85

Vertebrates, 28
Volcanoes, 46

Wales, 24, 27, 28, 31
Wallace, Alfred Russell, 19
Warm-blooded, 85